THE PYGMY PLANET

JOHN STEWART WILLIAMSON

dition et mise en page : Culturea (Hérault, 34)
etrouvez notre catalogue sur http://culturea.fr/
ontact : infos@culturea.fr
nprimé en Allemagne par Books on Demand Gmbh
esign typographique et layout : Derek Murphy et Reedsy
SBN : 9791041985562
ate de parution : février 2024

"Nothing ever happens to me!" Larry Manahan grumbled under his breath, sitting behind his desk at the advertising agency which employed his services in return for the consideration of fifty a week. "All the adventure I know is what I see in the movies, or read about in magazines. What wouldn't I give for a slice of real life!"

Unconsciously, he tensed the muscles of his six feet of lean, hard body. His crisp, flame-colored hair seemed to bristle; his blue eyes blazed. He clenched a brown hammer of a fist.

Larry felt himself an energetic, red-blooded square peg, badly afflicted with the urge for adventure, miserably wedged in a round hole.

It is one of the misfortunes of our civilization that a young man who, for example, might have been an excellent pirate a couple of centuries ago, must be kept chained to a desk. And that seemed to be Larry's fate.

"Things happen to other people," he muttered.

"Why couldn't an adventure come to me?"

He sat, staring wistfully at a picture of a majestic mountain landscape, soon to be used in the advertising of a railway company whose publicity was handled by his agency, when the jangle of the telephone roused him with a start.

"Oh, Larry— " came a breathless, quivering voice.

Then, with a click, the connection was broken.

The voice had been feminine and had carried a familiar ring. Larry tried to place it, as he listened at the receiver and attempted to get the broken connection restored.

"Your party hung up, and won't answer," the operator informed him.

He replaced the receiver on the hook, still seeking to follow the thin thread of memory given him by the familiar note in that eager excited voice. If only the girl had spoken a few more words!

Then it came to him.

"Agnes Sterling!" he exclaimed aloud.

Agnes Sterling was a slender, elfish, dark-haired girl—lovely, he had thought her, on the occasions of their few brief meetings. Larry knew her as the secretary and laboratory assistant of Dr. Travis Whiting, a retired college professor known for his work on the structure of the atom. Larry had called at the home-laboratory of the savant, months before, to check certain statistics to be used for advertising purposes and had met the girl there. Only a few times since had he seen her.

Now she had called him in a voice that fairly trembled with excitement—and, he thought, dread! And she had been interrupted before she had time to give him any message.

For a few seconds Larry stared at the telephone. Then he rose abruptly to his feet, crammed his hat on his head, and started for the door.

"The way to find adventure is to go after it," he murmured. "And this is the invitation!"

It was not many minutes later that he sprang out of a taxi at the front of the building in which Dr. Travis Whiting made his home and maintained a private experimental laboratory.

It was a two-story stucco house, rather out of date, set well back from the sidewalk, with a scrap of lawn and a few straggling shrubs before it. The door was closed, the windows curtained blankly. The place seemed deserted and forbidding.

Larry ran up the uneven brick walk to the door and rang the bell. Impatiently, he waited a few moments.

No sound came from within. He felt something ominous, fateful, about the silent mystery that seemed to shroud the old house. For the first time, it occurred to him that Agnes might be in physical danger, as a result of some incautious experiment on the part of Dr. Whiting.

Instinctively, his hand sought the door knob. To his surprise, the door was unlocked. It swung open before him. For a moment he stared, hesitating, into the dark hall revealed beyond. Then, driven by the thought that Agnes might be in danger, he advanced impulsively.

The several doors opening into the hall were closed. The one at the back, he knew, gave admittance to the laboratory. Impelled by some vague premonition, he hastened toward it down the long hall and threw it open.

As he stepped inside the room, his foot slipped on a spot of something red. Recovering his balance with difficulty, he peered about.

Bending down, Larry briefly examined the red spot on which he had slipped. It was a pool of fresh blood which had not yet darkened. Lying beside it, crimson-splashed, was a revolver. As he picked up the weapon, he cried out in astonishment.

Something had happened to the gun. The trigger guard was torn from it, and the cylinder crushed as if in some resistless grasp; the stock was twisted, and the barrel bent almost into a circle. The revolver had been crumpled by some terrific force—as a soft clay model of it might have been broken by the pressure of a man's hand.

"Crimson shades of Caesar!" he muttered, and dropped the crushed weapon to the floor again.

His eyes swept the silent laboratory.

It was a huge room, taking up all the rear part of the house, from the first floor to the roof. Gray daylight streamed through a sky-light, twenty feet overhead. The ends of the vast room were cluttered with electrical and chemical apparatus; but Larry's eye was caught at once by a strange and complex device, which loomed across from him, in the center of the floor.

Two pillars of intense light, a ray of crimson flame and another of deeply violet radiance, beat straight down from a complicated array of enormous, oddly shaped electron tubes, of mirrors and lenses and prisms, of coils and whirling disks, which reached almost to the roof. Upright, a yard in diameter and almost a yard apart, the strange columns of light were sharp-edged as two transparent cylinders filled with liquid light of ruby and of amethyst. Each ray poured down upon a circular platform of glass or polished crystal.

Hanging between those motionless cylinders of red and violet light was a strange-looking, greenish globe. A round ball, nearly a yard in diameter, hung between the rays, almost touching them. Its surface was oddly splotched with darker and lighter areas. It was spinning steadily, at a low rate of speed. Larry did not see what held it up; it seemed hanging free, several feet above the crystal platforms.

Reluctantly he withdrew his eyes from the mysterious sphere and looked about the room once more. No, the laboratory was vacant of human occupants. No one was hidden among the benches that were cluttered with beakers and test tubes and stills, or among the dynamos and transformers in the other end of the room.

A confusion of questions beat through Larry's brain.

What danger could be haunting this quiet laboratory? Was this the blood of Agnes Sterling or the scientist who employed her that was now clotting on the floor? What terrific force had crumpled up the revolver? What had become of Agnes and Dr. Whiting? And of whatever had attacked them? Had Agnes called him after the attack, or before?

Despite himself, his attention was drawn back to the little globe spinning so regularly, floating in the air between the pillars of red and violet flame. Floating alone, like a little world in space, without a visible support, it might be held up by magnetic attraction, he thought.

A tiny planet!

His mind quickened at the idea, and he half forgot the weird mystery gathering about him. He stepped nearer the sphere. It was curiously like a miniature world. The irregular bluish areas would be seas; the green and the brown spaces land. In some parts, the surface appeared mistily obscured—perhaps, by masses of cloud.

Larry saw an odd-looking lamp, set perhaps ten feet behind the slowly spinning, floating ball, throwing upon it a bright ray of vividly blue light. Half the strange sphere was brilliantly illuminated by it; the rest was in comparative darkness. That blue lamp, it came to Larry, lit the sphere as the sun lights the earth.

"Nonsense!" he muttered. "It's impossible!"

Aroused by the seeming wonder of it, he was drawn nearer the ball. It spun rather slowly, Larry noted, and each rotation consumed several seconds. He could distinguish green patches that might be forests, and thin, silvery lines that looked like rivers, and broad, red-brown areas that must be deserts, and the broad blue stretches that suggested oceans.

"A toy world!" he cried. "A laboratory planet! What an experiment— "

Then his eyes, looking up, caught the glistening, polished lens of a powerful magnifying glass which hung by a black ribbon from a hook on one of the heavy steel beams which supported the huge mass of silently whirring apparatus.

Eagerly, he unfastened the magnifier. Holding it before his eyes, he bent toward the strange sphere spinning steadily in the air.

"Suffering shades of Caesar!" he ejaculated.

Beneath the lens a world was racing. He could see masses of vividly green forest; vast expanses of bare, cracked, ocherous desert; wastes of smooth blue ocean.

Then he was gazing at—a city?

Larry could not be sure that he had seen correctly. It had slipped very swiftly beneath his lens. But he had a momentary impression of tiny, fantastic buildings, clustered in an elflike city.

A pygmy planet, spinning in the laboratory like a world in the gulf of space! What could it mean? Could it be connected with the strange call from Agnes, with the blood on the floor, with the strange and ominous silence that shrouded the deserted room?

"Oh, Larry!" a clear, familiar voice rang suddenly from the door. "You came!"

Startled, Larry leaped back from the tiny, whirling globe and turned to the door. A girl had come silently into the room. It was Agnes Sterling. Her dark hair was tangled. Her small face was flushed, and her brown eyes were wide with fear! In a white hand, which shook a little, she carried a small, gold-plated automatic pistol.

She ran nervously across the wide floor to Larry, with relief dawning in her eyes.

"I'm so glad you came!" she gasped, panting with excitement. "I started to call you on the phone, but then I was afraid it would kill you if you came! Please be careful! It may come back, any minute! You'd better go away! It just took Dr. Whiting!"

"Wait a minute," Larry put in. "Just one thing at a time. Let's get this straight. To begin with, what is it that might kill me, and that got the doctor?"

"It's terrible!" she gasped, trembling. "A monster! You must go away before it comes back!"

Larry drew a tall stool from beside one of the crowded tables and placed it beside her.

"Don't get excited," he urged. "I'm sure everything will be all right. Just sit down, and tell me about it. The whole story. Just what is going on here, and what happened to Dr. Whiting."

He helped her upon the stool. She looked up at him gratefully, and began to speak in a rapid voice.

"You see that little planet? The monster came from that and carried the doctor back there. And I know it will soon be back for another victim—for sacrifice!"

She had pointed across the great room, toward the strange little globe which hung between the pillars of red and violet light.

"Please go slow!" Larry broke in. "You're too fast for me. Are you trying to tell me that that spinning ball is really a planet?"

Agnes seemed a little more composed, though she was still flushed and breathing rapidly. Her small hand still gripped the bright automatic.

"Yes, it is a planet. The Pygmy Planet, Dr. Whiting called it. He said it was the great experiment of the century. You see, he was testing evolution. We began with the planet, young and hot, and watched it until it is now almost as old as Mars. We watched the change and development of life upon it. And the rise and decay of a strange civilization. Until now its people are strange things, with human brains in mechanical bodies, worshiping a rusty machine like a god— "

"Go slow!" Larry pleaded again. "I don't see—Did the doctor build—create—that planet himself?"

"Yes. It began with his work on atomic structure. He discovered that certain frequencies of the X-ray—so powerful that they are almost akin to the cosmic ray—have the power of altering electronic orbits. Every atom, you know, is a sort of solar system, with electrons revolving about a proton.

"And these rays would cause the electrons to fall into incredibly smaller orbits, causing vast reduction in the size of the atoms, and in the size of any object which the atoms formed. They would cause anything, living or dead, to shrink to inconceivably microscopic dimensions—or restore it to its former size, depending upon the exact wave-length used.

"And time passes far more swiftly for the tiny objects—probably because the electrons move faster in their smaller orbits. That is what suggested to Dr. Whiting that he would be able to watch the entire life of a planet, in the laboratory. And so, at first, we experimented merely with solitary specimens or colonies of animals.

"But on the Pygmy Planet, we have watched the life of a world—the whole panorama of evolution—"

"It seems too wonderful!" Larry muttered. "Could Dr. Whiting actually decrease his size and become a dwarf?"

"No trick at all," Agnes assured him. "All you have to do is stand in the violet beam, to shrink. And move over in the red one, when you want to grow. I have been several times with Dr. Whiting to the Pygmy Planet."

"Been— " Larry stopped, breathless with astonishment.

"See the little airplane," Agnes said, pointing under the table.

Larry gasped.

Beneath the table stood a toy airplane. The spread of its glistening, perfect wings was hardly three feet. A wonderful, delicate toy, accurate in every detail of propeller, motor and landing gear, of brace and rudder and aileron. Then he realized that it was no toy at all, but a faithful miniature of a commercial plane. A complete, tiny copy of one of the latest single-motor, cabin monoplane models.

"It looks like it would fly," he said "a friend of mine his a big one, just like it! Taught me to fly it, last summer vacation. This is the very image of it!"

"It will fly!" Agnes assured him, now composed enough to smile at his amazement. "I have been with the doctor to the Pygmy Planet in it.

"You stand in the violet ray until you're about three inches high," she explained, "and then get into the plane. Then you fly up and into the violet ray at the point where it touches the planet, and remain there while you grow smaller. When you are the right size, all you have to do is drop to the surface, and land. To come away, you rise into the red ray and stay in it till you grow to proper size, when you come down and land."

"You—you've actually done that?" he gasped. "It sounds like a fairy story!"

"Yes, I've done it," she assured him. Then she shuddered apprehensively. "And the things—the machine-monsters, Dr. Whiting called them—have learned to do it, too. One of them came down the red ray, and attacked him. The doctor had a gun—but what could he do against one of those?" She shivered.

"It carried him back up the violet beam. Just a few minutes ago, I started to phone you. Then I was afraid you would be hurt— "

"Me, hurt?" Larry burst out. "What about you, here alone?"

"It was my business. Dr. Whiting told me there might be danger, when he hired me."

"And now, what can we do?" Larry demanded.

"I don't know," she said slowly. "I'm afraid one of the monsters will be back after a new victim. We could smash the apparatus, but it is too wonderful to be destroyed. And besides, Dr. Whiting may have escaped. He may be alive there, in the deserts!"

"We might fly up, in the little plane," Larry proposed, doubtfully. "I think I could pilot it. If you want— "

The girl's body stiffened. Her brown eyes widened with sudden dread, and her small face went pale. She slipped quickly from the stool, drawing in her breath with a sort of gasp. The hand that gripped the automatic trembled a little.

"What's the matter?" Larry cried.

"I thought— " she gasped, "I think I see something in the ray! The machine-monster is coming back!"

Her lips tightened. She lifted the little automatic and began to shoot into the pillar of crimson fire beside the tiny, spinning globe.

Larry, watching tensely, saw a curious, bird-like something fluttering about in the red ray, swiftly growing larger!

Deliberately, and pausing to aim carefully for each shot, the girl emptied the little gun at the figure. Her body was rigid, her small face was firmly set, though she was breathing very fast.

A curious numbness had come over Larry. His only physical sensations were the quick hammering of his heart, and a parching dryness in his throat. Terror stiffened him. Though he would not have admitted it, he was paralyzed with fear.

The glittering thing that fluttered about in the crimson ray was not an easy target. When the gun was empty, it seemed still unharmed. And its wings had increased to a span of a foot.

"Too late!" Agnes gasped. "Why didn't we do something?"

Trembling, horror-stricken, she shrank toward Larry.

He was staring at the thing in the pillar of scarlet light.

It had dropped to the crystal disk upon which the red ray fell from the huge, glowing tube above. It stood there, motionless except for the swift increase of its size.

Larry gazed at it, lost in fear and wonder. It was like nothing he had ever seen. What was it that Agnes had said, of machine-monsters, of human brains in mechanical bodies? His brain reeled. He strained his eyes to distinguish the monstrosity more clearly. It was veiled in crimson flame; he could not see it distinctly.

But suddenly, when it was as tall as himself, it sprang out into the room, toward Larry and the shuddering girl. Just off the crystal disk, beyond the scarlet pillar of fire, it paused for long seconds, seeming to regard them with malevolent eyes.

For the first time, Larry could see it plainly.

Its body, or its central part, was a tube of transparent crystal; an upright cylinder, rounded at upper and lower ends. It was nearly a foot in diameter, and four feet long. It seemed filled with a luminous, purple liquid.

About the cylinder were three bands of greenish, glistening metal. Attached to the lower band were four jointed legs of the same bright green metal, upon which the strange thing stood.

Set in the middle band were two glittering, polished lenses, which seemed to serve as eyes, and Larry felt that they were gazing at him with malevolent menace. Behind the eyes, two wings sprang from the green band. Ingenious, folding wings, of thin plates and bars of green metal.

And from the upper band sprang four slender, glistening, whip-like tentacles, metallic and brilliantly green, two yards in length. They writhed with strange life!

It seemed a long time to Larry that the thing stood, motionless, seeming to stare evilly at them with eye-like lenses. Then, lurching forward a little, it moved toward them upon legs of green metal. And now Larry saw another amazing thing about it.

Floating in the brilliant violet liquid that filled the crystal tube was a gray mass, wrinkled and corrugated. This was divided by deep clefts into right and left hemispheres, which, in turn were separated into larger upper and smaller lower segments. White filaments ran through the violet liquid from its base toward the three rings or bands of green metal that encircled the cylinder.

In an instant, Larry realized that the gray mass was a human brain. The larger, upper part the cerebrum, the smaller mass at the back the cerebellum. And the white filaments were nerves, by means of which this brain controlled its astounding, mechanical body!

A brain in a machine!

The violet liquid, it came to Larry in his trance of wonder, must take the place of blood, feeding the brain-cells, absorbing waste.

An eternal mind, within a machine! Free from the ills and weaknesses of the body. And devoid, too, of any pity, of any tender feelings. A cold and selfish mind, without emotion—unless it might worship itself or its mechanical body.

It was this monster that had spilt the pool of blood drying on the floor, near the door. And it was these glistening, green, snake-like tentacles that had crumpled the revolver into a broken mass of steel!

Abruptly the machine-monster darted forward, running swiftly upon its four legs of green metal. Slender tentacles reached out toward the shuddering girl at Larry's shoulder.

"Run!" Agnes gasped to him, quickly. "It will kill you!"

The girl tried to push him back.

As she touched him, Larry recovered from his daze of wondering fear. Agnes was in frightful danger, and facing it with quiet courage. He must find a weapon!

Wildly, he looked about him. His eyes fell upon the tall, heavy wooden stool, upon which Agnes had been sitting.

"Get back!" he shouted to her.

He snatched up the stool, and, swinging it over his head, sprang toward the machine of violet-filled crystal and glittering green metal.

"Stop!" Agnes screamed, in a terrified voice. "You can't— "

She had run before him. He seized her arm and swung her back behind him. Then he advanced warily toward the machine-monster, which had paused and seemed to be regarding him with sinister intentness, through its glistening crystal eye-lenses.

With all his strength, Larry struck at the crystal cylinder, swinging the stool like an ax. A slender, metallic green tentacle whipped out, tore the stool from his hands, and sent it crashing across the room, to splinter into fragments on the opposite wall.

Larry, sent off his balance, staggered toward the glittering machine. As he stumbled against the transparent tube that contained the brain, he clenched his fist to strike futilely at it.

A snake-like metal tentacle wrapped itself about him; he was hurled to the floor, to sprawl grotesquely among broken apparatus.

His head came against the leg of a bench. For a few moments he was dazed. But it seemed only a few seconds to him before he had staggered to his feet, rubbing his bruised head. Anxiously, he peered about the room.

The machine-monster and Agnes were gone!

He stumbled back to the mass of apparatus in the center of the huge laboratory. Intently, he gazed into the upright pillar of crimson flame. Nothing was visible there.

"No, the other!" he gasped. "The violet is the way they went."

"No, the other!" he gasped. "The violet is the way they went."

He turned to the companion ray of violet radiance that beat straight down on the opposite side of the tiny, whirling planet. And in that motionless torrent of chill violet flame he saw them.

Tiny, already, and swiftly dwindling!

With green wings outspread, the machine-monster was beating swiftly upward through the pillar of purple-blue flame. And close against the crystal tube that contained its brain, was Agnes, held fast by the whip-like tentacles of glistening green metal.

Larry moved to spring after them, into the torrent of violet light. But sudden caution restrained him.

"I'd shrink, too!" he muttered. "And then where would I be? I'd be standing on the glass platform, I guess. And the thing flying off over my head!"

He gazed at the rapidly dwindling forms of Agnes Sterling and her amazing abductor. As it grew smaller, the machine-monster flew higher in the violet beam, until it was opposite the tiny, spinning planet.

The distance between the red and the violet rays was just slightly more than the diameter of the pygmy world. The sphere hung between them, one side of it a fraction of an inch from the red, the other as near the violet.

Opposite the elfin planet, the monster ceased to climb. It hung there in the violet ray, an inch from the surface of the little world.

And still it swiftly dwindled. It was no larger than a fly, and Larry could barely distinguish the form of the girl, helpless in the green tentacles.

Soon she and the monster became a mere greenish speck.... Suddenly they were gone.

For a little time he stood watching the point where they had vanished, watching the red and the violet rays that poured straight down upon the crystal disks, watching the tiny, green-blue planet spinning so steadily between the bright rays.

Abruptly, he recovered from his fascination of wonder.

"What did she say?" he muttered. "Something about the monsters carrying off people to sacrifice to a rusty machine that they worship as a god! It took her—for that!"

He clenched his fists; his lips became a straight line of determination.

"Then I guess we try a voyage in the little plane. A slim chance, maybe. But decidedly better than none!"

He returned to the table, dropped on his knees, inspected the tiny airplane. A perfect miniature, delicately beautiful; its slim, small wings were bright as silver foil. Carefully, he opened the door and peered into the diminutive cabin. Two minute rifles, several Lilliputian pistols, and boxes of ammunition to match, lay on the rear seat of the plane.

"So we are prepared for war," he remarked, grinning in satisfaction. "And the next trick, I suppose, is to get shrunk to fit the plane. About three inches, she said. Lord, it's a queer thing to think about!"

He got to his feet, walked back to the machine in the center of the room, with its twin pillars of red and violet flame, and the tiny world floating between them. He started to step into the violet ray, then hesitated, shivering involuntarily, like a swimmer about to dive into icy cold water.

Turning back to one of the benches, he picked up a wooden funnel-rack, and tossed it to the crystal disk beneath the violet ray. Slowly it decreased in size, until it had vanished from sight.

"Safe, I suppose," he muttered. "But how do I know when I'm small enough?"

After a moment he picked up a glass bottle which measured about three inches in height, set it on the floor, beside the crystal disk.

"I dive out when I get to be the size of the bottle," he murmured.

With that, he leaped into the violet beam.

He felt no unusual sensation, except one of pleasant, tingling warmth, as if the direct rays of the sun were bearing down upon him. For a moment he feared that his size was not being affected. Then he noticed, not that he appeared to become smaller, but that the laboratory seemed to be growing immensely larger.

The walls seemed to race away from him. The green-blue sphere of the tiny planet which he proposed to visit expanded and drew away above his head.

Abruptly fearful, alarmed at the hugeness of the room, he turned to look at the bottle he had placed to serve as a standard of size. It had grown with everything else, until it seemed to be about three feet high.

And it was swiftly expanding. It reached to the level of his shoulder. And higher!

He ran to the edge of the crystal disk, which now seemed a floor many yards across, and leaped from its edge. It was a dozen steps to where he had left the bottle. And it was as tall as himself!

He started across the floor of the laboratory toward the table under which the toy plane stood. The incredible immensity of his surroundings awed him strangely. The walls of the room seemed distant, Cyclopean cliffs; the roof was like a sky. Table legs towered up like enormous columns.

It seemed a hundred yards across the strangely rough floor to the plane. As he drew near it, it gave him huge satisfaction to see that it was of normal size, correctly proportioned to his own dimensions.

"Great luck," he muttered, "that I can fly!"

He paused, as he reached the cabin's open door, to wonder at the astounding fact that a little while ago he had opened that door with a hand larger than his entire body now was.

"I guess this is my day of wonders!" he muttered. "Allah knows I had to wait long enough for it!"

First he examined the weapons in the cabin. There were two heavy sporting rifles and two .45 automatics. There were also two smaller automatics, which, he supposed, had been intended for Agnes' use. And there was abundant ammunition.

Then he inspected the plane. It looked to be in excellent condition in every way. The gasoline and oil tanks were full.

He set about starting the motor, using the plane's inertia starter, which was driven by an electric motor. Soon the engine coughed, sputtered, and gave rise to a roaring, rhythmic note that Larry found musical.

When the motor was warm, he opened the throttle and taxied out from beneath the colossal table, and across the laboratory floor toward the Titanic mechanism in the center of the room. The disk of crystal was set almost flush with the floor, its edge beveled. The plane rolled easily upon it, and out into the Cyclopean pillar of violet flame.

Once more, Larry felt the sensation that everything about him except the plane itself, was expanding inconceivably in size. Soon the laboratory's walls and roof were lost in hazy blue distance. He could distinguish only the broad, bright field formed by the surface of the crystal disk, with the floor stretching away beyond it like a vast plain. And above, the green-blue sphere of the tiny planet, bright on one side and dark on the other, so that it looked like a half-moon, immensely far-off.

As he waited, he noticed a curious little dial, in a lower corner of the instrument board, which he had not seen at first. One end of its graduated scale was marked, "Earth Normal," the other, "Pygmy Planet Normal." A tiny black needle was creeping slowly across the scale, toward "Pygmy Planet Normal."

"That's how we tell what size we are without having to look at a bottle," he muttered.

When the area of the crystal platform appeared to be about half a square mile, he decided that he would now have sufficient space to spiral up the violet ray toward the planet. If he waited too long to start, the distance would become impossibly great.

He gave the little plane the gun. The motor thundered a throbbing song; the ship rolled smoothly forward over the polished surface, gained flying speed and took the air without a shock.

"Feels good to hold the stick again!" Larry murmured.

Making small circles to keep within the upright pillar of violet radiance, he climbed steadily and as rapidly as possible, keeping his eyes upon the brilliant half-moon of the Pygmy Planet.

The strangest flight in the annals of aviation! He was flying toward a goal that, a few minutes before, he could have touched. Toward a goal that, at the beginning of his flight, was only a few lengths of his plane away. And his size dwindled so rapidly as he flew that the planet seemed to swell and draw away from him.

As Larry and the plane grew smaller, the relative size of the violet ray increased, so there was no longer much danger of flying out of it. It seemed that he flew through a world of violet flame.

He met a curious problem in time. It is evident that time passes faster for a small animal than for a large one, because nerve currents require a shorter time in transit, and all thought and action is consequently speeded up. It took a hundred-foot dinosaur nearly a second to know that his tail had been pinched. A fly can get under way in time to escape a descending swatter. The Pygmy Planet rotated in a few seconds of earth time; one of its inhabitants might have lived, aged, and died in the duration of a single day in our larger world.

So Larry found that time seemed to pass more rapidly, or rather that the time of the world he had left appeared to move more slowly, as he adventured into smallness. He had been flying, it seemed to him, nearly an hour when he reached the level of the planet's equator.

Now it seemed a vast world, filling half the visible universe. He flew toward it steadily, until he knew, by the fading before him of the violet flame which now seemed to fill all space, that he was near the edge of the ray. And as he flew, he watched the little scale, upon which the black needle was now nearing the line marked, "Pygmy Planet Normal."

Circling slowly, keeping always on the level of the planet's equator, and near the edge of the violet ray, so as to be as close as possible to his landing place when he reached the proper size, he watched the creeping black needle.

Too, he scanned with eager eyes the planet floating before him. Bare, red deserts; narrow strips of green vegetation; shrunken, blue oceans; silvery lines of rivers, passed in fascinating panorama beneath his eyes. The rate of the planet's spinning seemed continually to lessen, with the changing of his own sense of time.

Agnes! Larry thought of her with a curious, eager pain in his heart. She was somewhere on that strange, ancient world, a prisoner of weird machine-monsters! Intended victim of a grotesque sacrificial ceremony!

Could he find her, in the vastness of an unfamiliar world? And having found her, would there be a chance to rescue her from her hideous captors? The project seemed insane. But Larry felt a queer, unfamiliar urge, which, he knew, would drive him on until he had discovered and saved her—or until he was dead.

At last, when it seemed to Larry nearly three hours since he had begun this amazing flight, the crawling ebon needle reached the mark, "Pygmy Planet Normal."

He flew out of the wall of violet flame toward the planet's surface. Before, the distance between the planet and the ray's edge had seemed only the fraction of an inch. Now it appeared to be many miles.

Abruptly the Pygmy Planet, which had seemed to be beside him, appeared to swing about, so that it was beneath him. He knew that it was a change merely in his sensations. He was feeling the gravitation of the new world. It was pulling him toward it!

He cut the throttle, and settled the plane into a long glide, a glide that was to end upon the surface of a new planet!

In what seemed half an hour more, Larry had made a safe landing upon the Pygmy Planet.

He had come down upon a stretch of fairly smooth, red, sandy desert, which seemed to stretch illimitably toward the rising sun, which direction Larry instinctively termed "east."

To the "west" was a line of dull green—evidently the vegetation along a stream. The ocher desert was scattered with sparse clumps of reddish, spiky scrub. Larry taxied the plane into one of those thickets. Finding canvas and rope in the cabin, he staked down the machine, and muffled the motor.

Then, selecting a rifle and a heavy automatic from the weapons in the cabin, and filling his pockets with extra ammunition, he left the plane and set out with brisk steps toward the green line of vegetation.

"I'll follow along the river," he reasoned. "It may lead me somewhere and it will show the way back to the plane.

I may come across something in the way of a clue. Can't go exploring by air, or I'll burn up all the gas and be stranded here!"

To his surprise, the water course proved to be an ancient canal, walled with crumbling masonry. Its channel was choked with mud and thorny, thick-leaved desert shrubs of unfamiliar variety; but a feeble current still flowed along it.

After some reflection, Larry set out along the banks of the canal.

He followed it for two days.

Curious straight bars of light were visible across the sky—a band of violet in the morning; one of crimson at evening. Their apparent motion was in the same direction as that of the sun. The bars of light puzzled him considerably before it occurred to him that they must be the red and violet rays.

"So you wait till evening, and then fly up into the red ray, to go home," he muttered. "But I may not need that information," he added grimly. "Seems to be a pretty big job to search a planet on foot, for one person. And I'm not going back without Agnes!"

In the afternoon of the second day, he came within view of a city. He could discern vast, imposing walls and towers of dark stone. It stood in the barren red desert, far back from the green line of the old canal. Larry left the canal and started wearily across toward it. He had covered several miles of the distance before he saw that the lofty towers were falling, the magnificent walls crumbling. The city was ruined, dead, deserted!

The realization brought him a great flood of despair. He had hoped to find people—friends, from whom he might get food, and information about this unfamiliar planet. But the city was dead.

Larry was standing there, in the midst of the vast red plain between ruined city and ruined canal. Tired, hungry, lonely and hopeless. He was looking up at the white "sun," trying to comfort himself with the thought that the brilliant luminary was merely a queer blue lamp, that he was upon a tiny experimental world in a laboratory. But the thought brought him no relief; only confusion and a sense of incredulity.

Then he saw the machine-monster.

A glittering, winged thing of crystal and green metal, identical with the one he had encountered in the laboratory. It must already have seen him, for it was dropping swiftly toward him.

Larry started to run, took a few staggering steps. Then he recalled the heavy rifle slung over his shoulder. Moving with desperate haste, he got it into his hands and raised it just as the monster dropped to the red sand a dozen yards away from him.

Steadily he covered the crystal cylinder within which the thing's brain floated in luminous violet liquid. His finger tightened on the trigger, ready to send a heavy bullet crashing into it. Then he paused, swore softly, lowered the gun.

"If I kill it," he murmured, "I may never find Agnes. And if I let it carry me off, it may take me where she is."

He walked toward the monster, across the red sand.

It stood uncertainly upon green metal legs, seeming to stare at him strangely with eye-like lenses. Its wings of thin green metal plates, were folded; its four green tentacles were twitching oddly.

Abruptly, it sprang upon him.

A green tentacle seized the rifle and snatched it from his hands. He felt the automatic pistol and the ammunition being removed from his pockets.

Then, firmly held in the flexible arms of green metal, he was lifted against the cylinder of violet liquid. The monster spread its broad emerald wings, and Larry was swiftly borne into the air.

In a few moments the wide ruins of the ancient city were spread below, with the green line of the choked canal cutting the infinite red waste of the desert beyond it.

The monster flew westward.

The monster flew westward.

For a considerable time, nothing save barren, ocherous desert was in view. Then Larry's weird captor flew near a strange city. A city of green metal. The buildings were most fantastic—pyramids of green, crowned with enormous, glistening spheres of emerald metal. An impassable wall surrounding the city.

Larry had expected the monster to drop into the city. But it carried him on, and finally settled to the ground several miles beyond. The green tentacles released him, as the thing landed, and he sprawled beside it, dizzy after his strange flight.

As Larry staggered uncertainly to his feet, he saw that the monster had released him in an open pen. It was a square area, nearly fifty yards on each side, and fenced with thin posts or rods of green metal, perhaps twenty feet high. Set very close together, and sharply pointed at the top, they formed a barrier apparently insurmountable.

In the center of the pen was a huge and strange machine, built of green metal. It looked very worn and ancient; it was covered with patches of bluish rust or corrosion. At first it looked quite strange to Larry; then he was struck by a vaguely familiar quality about it. Looking closer, he realized that it was a colossal steam hammer!

Its design, of course, was unfamiliar. But in the vast, corroded frame he quickly picked out a steam chest, cylinder, and the great hammer, weighing many tons.

He gasped when his eyes went to the anvil.

A man was chained across it.

A man in torn, grimy clothing, fastened with fetters of green metal upon wrists and ankles, so that his body was stretched beneath the massive hammer. He seemed to be unconscious; upon his head, which was turned toward Larry, was a red and swollen bruise.

The monster which had dropped Larry within the pen rose again into the air. And Larry started forward, trying to remember just what Agnes had told him of a machine to which the monsters sacrificed.

This must be the machine—this ancient steam hammer!

As he moved forward, Agnes came into view.

She walked around the massive base of the great machine, carrying a bowl filled with a fragrant brown liquid. She stopped at sight of Larry, and uttered a little cry. The bowl fell from her hands, and the fragrant liquid splashed out on the ground. Her brown eyes went wide with delighted surprise; then a look of pain came into them.

"Larry, Larry!" she cried. "Why did you come?"

"To get you," he answered, trying to speak as lightly as he could. "And the best way I knew to find you was to let one of the monsters bring me. Cheer up!" But even to himself, his voice had a tone of discouragement.

She smiled wanly. "I don't see anything to be cheerful about." Her small face was set and a little white. "Dr. Whiting is going to be smashed under the hammer of this dreadful machine, whenever the steam is up. Then it is my turn. And yours. That's nothing to laugh about."

"But we aren't smashed yet!" Larry insisted.

"By the way, what was that in the bowl?" he went on, glancing down. "I forgot to bring lunch." He grinned.

She looked down, startled.

"Oh. Dr. Whiting's soup. Poor fellow, I'm afraid he'll never awake to eat it. There's plenty more. Come around here."

She picked up the bowl and led him around the base of the machine; then she filled the bowl again with the fragrant, red-brown liquid, from a tall urn of green metal. Larry took the dish eagerly and gulped down the rather insipid and tasteless food.

"And the monsters worship this old steam hammer?" he inquired, when his hunger was appeased.

"Yes. I think the thing is worked by steam generated by volcanic heat. Anyhow, there isn't any boiler, and the steam pipe comes up out of the ground. You can see that. So it runs on, without any attention—though I guess the heat is dying down, since it is several days between blows of the hammer.

"And I guess the monsters have forgotten how they used to rule machines. They seem to have depended upon machines, even giving up their own bodies for mechanical ones, until the machine rules them.

"And when this old hammer kept pounding on through the ages, using volcanic steam, I guess they got to considering it alive. They began to regard it as a sort of god. And when they got the idea of giving it sacrifices, it was natural enough to place the victims under the hammer."

They went back to Dr. Whiting who was chained across the anvil. He was still breathing, but unconscious. He had been injured in a struggle with the monsters, and his body was much emaciated. Agnes explained that he had been a prisoner in the pen for many months of the time of this world, waiting his turn to die; she said that the monsters had just completed the extermination of another race upon the Pygmy Planet, and were just turning to the greater world for victims.

Larry noticed that the great hammer was slowly rising in its guides, as the pressure of the steam from the planet's interior increased. In a few hours—just at sunset—it reached the top of its stroke.

The air above the pen was suddenly filled with glittering swarms of the green-winged monsters, sweeping slowly about, in measured flight, with strange order in their masses. They had come to witness the sacrifice!

With an explosive rush of steam, the hammer came down!

The ground trembled beneath the terrific blow; the roaring of escaping steam and the crash of the impact were almost deafening. A heavy white cloud shrouded the corroded green machine.

When the hammer slowly lifted, only a red smear was left....

Agnes had shrunk, trembling, against Larry's shoulder. He had put his arms about her and was holding her almost fiercely.

"My turn next," she whispered. "And don't try to fight them. It will only make them hurt you!"

"I can't let them take you, Agnes!" Larry cried, in an agonized tone. And the words seemed to leap out, of themselves, "Because I love you!"

"You do?" Agnes cried, in a thin, choking voice, pressing herself against him. "Ever since the first time you came to the laboratory—"

A score of the monster forms of violet-filled crystal and gleaming green metal had dropped into the pen. They tore Agnes from Larry's arms, hurling him roughly to the ground, at the bottom of the green metal fence. For some time he was unconscious.

When he had staggered painfully to his feet, it was night. The monsters were gone; the starless sky was black and empty. Calling out weakly, and stumbling about the pen, he found Agnes. She was chained where Dr. Whiting had been.

She was conscious, unharmed. For a time they talked a little, exchanging broken, incoherent phrases. Then they went to sleep, lying on the anvil, beneath that mighty hammer that was slowly lifting to strike another fearful blow.

When the "sun" had risen again, Larry brought Agnes some of the brown soup from the metal urn, which had been filled again. Then, when he had satisfied himself, he started clambering up the massive frame of the hammer.

If he could put it out of commission!

It was a difficult task. He slipped back many times, and finally had to choose another place to make the ascent.

Twice he slipped and almost fell from a considerable height. But finally he reached the massive wheel of the valve which seemed to control the admission of steam into the cylinder above the hammer.

If he could but close that, the steam would be confined in the chest below. And when the pressure reached a certain point, something should happen!

The valve was not easy to turn; it seemed fixed with the corrosion of ages. For hours Larry wrestled with it. Then he left it, realizing that he must find something to use for a hammer.

A vigorous search of the pen's hard earth floor failed to reveal any stone that would do. He turned his attention to the machine, and presently saw a slender projecting lever, high up on the side of the vast frame, which looked as if it had been weakened by corrosion.

After a perilous climb, he reached the bar of green metal and swung his weight upon it. It broke, and he plunged to the ground with the bar in his hands.

Clambering up once more to the great valve, he hammered it until the rust that stiffened it was loosened. Then he struggled with the valve until it was closed.

"We'll see what happens!" he muttered.

Returning to the ground, he set to work to break the green metal fetters upon Agnes' wrists and ankles, using the broken lever as hammer and file.

For the greater part of six days he toiled at that task, while the great hammer rose slowly. But the green metal seemed very hard. One arm was free at the end of the second day, the other on the fourth. He had one ankle loose on the morning of the sixth day. But as evening came on, and the great hammer reached the top of its stroke, the fourth chain still defied him.

Before sunset, a swarm of the monsters appeared, wheeling on green wings. He was forced to leave the work, hiding his improvised file.

Agnes still lay across the anvil, to conceal from the monsters the fact that the chains were broken. Larry sat close beside her, nursing hands that were blistered and sore from his days of filing at the chains.

A sudden clatter came from the huge mechanism above them, and a sharp hiss of steam, which became louder.

"It works!" Larry whispered to Agnes. "The old valve held, and the steam can't get into the cylinder to smash us! But Allah knows what will happen when the pressure rises in that old steam chest!"

Darkness came. Dusk swallowed the wheeling machine-monsters. All night Larry and Agnes waited silently, together on the great anvil, listening to the hissing of steam from above, which was slowly becoming a shrill monotonous scream; monotonous, always higher, shriller.

The "sun" rose again. Still the green-winged monsters wheeled about. They came in glittering swarms, thousands of them. They came nearer the machine now, and flew about more swiftly, is if excited.

Then it happened.

There was a roar like thunder, and a colossal, bellowing explosion. The air was filled suddenly with scalding steam, and with screaming fragments of the bursting steam chest. In the midst of it all, Larry felt a crushing blow upon the head. And a blanket of darkness fell upon him....

"The monsters are all gone, darling," Agnes' voice reached him. "As though they were very much frightened. And a piece of the old hammer hit the fence and knocked a hole in it. You must go. Leave me— "

"Leave you?" Larry groaned, struggling to sit up. "Not a bit of it!" He touched his head gingerly, felt a swollen bruise.

Collecting a few fragments of the wrecked machine, to serve as tools, he fell to work again upon Agnes' remaining chain. Already he had cut a deep groove in it. Two hours later, it was broken.

Carrying the metal urn of brownish liquid, they crept out through the hole in the fence, which had been torn by the flying fragment of a broken casting of green metal. They left the wreck of the machine which a strange race had worshiped as a bloody god and hurried furtively into the desert of red sand.

Making a wide circuit about the fantastic city of green metal, which Larry had seen from the air, they struck out eastward across the desolate ocherous waste. The food in the urn, eaten sparingly, lasted until the end of the eighth day.

On the morning of the ninth, they came in view of the green line of the ancient canal. It was hours later that they staggered weakly over its wall of crumbling masonry, clambered down into the muddy, weed-grown channel, and drank thirstily of green, tepid water.

Larry found his old trail, beyond the canal. They followed it back. In the middle of the afternoon they stumbled up to the thicket of spiky desert growth, in which Larry had hidden the plane.

The machine was undamaged.

The machine was undamaged.

Before sunset, Larry had removed the stake ropes, slipped the canvas cover from the motor, turned the plane around, inspected it, and examined the strip of smooth, hard red sand upon which he had landed.

Agnes pointed out the dim band of crimson across the sky, from north to south, slowly rising toward the zenith.

"That's the red ray," she said. "We fly into it."

"And a happy moment when we do," Larry rejoined.

He roused the motor to life.

As the bar of crimson light neared the zenith, the plane rolled forward across the sand and took off. Climbing steeply, Larry anxiously watched the approach of the red band. The gravitation of the Pygmy Planet seemed to diminish as he gained altitude, until presently he could fly vertically from it, without circling at all. He set the bow toward the scarlet bar across the sky before him.

And suddenly he was flying through ruby flame.

His eyes went to the little scale at the corner of the instrument board. He saw the little ebon needle waver, leave the mark designated "Pygmy Planet Normal" and start toward "Earth Normal."

For what seemed a long time, he was wheeling down the crimson ray. A few times he looked back at Agnes, in the rear seat. She had gone to sleep.

Then a vast, circular field was below—the crystal platform.

Larry landed the plane upon it, taxied to the center and stopped there, with the motor idling. The laboratory, taking shape in the blue abyss about him, seemed to contract swiftly.

Presently the plane covered most of the crystal disk. He taxied quickly off, stopped on the floor nearby, and cut the ignition. Agnes woke. Together they clambered from the plane's cabin and walked back into the crimson ray.

Once more the vast spaces of the room seemed to shrink, until it looked familiar once more. The Pygmy Planet, and the huge machine looming ever them, dwindled to natural size.

Agnes, watching a scale on the frame of the mechanism, which Larry had not noticed, leaped suddenly from the red ray, drawing him with her.

"We don't want to be giants!" she laughed.

Larry drew a deep breath, and looked about him. Once more he was in his own world, and surveying it in his normal size. He became aware of Agnes standing close against him. He suddenly took her in his arms and kissed her.

"Wait a minute," she objected, slipping quickly from his arms. "What are we going to do about the Pygmy Planet?

Those monsters might come again, even if you did wreck their god. And Dr. Whiting, poor fellow — But we mustn't let those monsters come back!"

Larry doubled up a brown fist and drove it with all his strength against the little globe that spun so steadily between the twin, upright cylinders of crimson and of violet flame. His hand went deep into it.

And it swung from its position, hung unsteadily a moment, and then crashed to the laboratory floor. It was crushed like a ball of soft brown mud. It spattered.

"Now I guess they won't come back," Agnes said. "A pity to spoil all Dr. Whiting's work, though."

Larry was standing motionless, holding up his fist and looking at it oddly. "I smashed a planet! Think of it. I smashed a planet! Just the other—why it was just this evening, at the office, I was wishing for something to happen!"